Australia's Environmental Issues

SALINITY

REDBACK publishing

Redback Publishing
PO Box 357 Frenchs Forest NSW 2086
Australia

www.redbackpublishing.com.au
orders@redbackpublishing.com.au

978-1-925860-30-6

Author: Peter Turner
Editor: Michael Anderson
Proofer: Marianne Lindsell
Designer: Redback Publishing

Original illustrations © Redback Publishing 2022
Originated by Redback Publishing

Reprinted 2022
Printed and bound in Malaysia

Acknowledgements
Abbreviations: l—left, r—right, b—bottom, t—top, c—centre, m—middle
We would like to thank the following for permission to reproduce photographs: (Images © shutterstock)

NATIONAL LIBRARY OF AUSTRALIA
A catalogue record for this book is available from the National Library of Australia

CONTENTS

INTRODUCTION: WHAT IS SALT?

Salt is a chemical that comes in many different forms, including sodium chloride. Sodium chloride is the most common form of edible salt, it is sold as table salt, rock salt and sea salt. Salt is an important part of our diet. Small amounts of salt are essential for the health and wellbeing of all living things. However, too much salt can be highly toxic and even deadly. Most plants, including Australia's food crops, are highly sensitive to salt.

Where Does Salt Come From?

In nature, salt is found in the oceans, rain, rocks and soil. Millions of years ago, parts of Australia were covered by an inland sea. As the water slowly disappeared, it left behind sediments that were high in salt. These sediments broke down and seeped into the soil, making much of the landscape naturally salty.

Australia is a large island surrounded by the sea, and consequently the rains that fall on Australia can also be high in salt. Winds pick up the salty water spray as they cross the oceans. The salt is carried high up into the atmosphere, where it becomes part of clouds. The salt is then released over Australia when it rains.

Rocks form over millions of years from layers of salt and other sediments that build up and are pressed together over time. As rocks erode, they release their salts, which seep into the soil and ground water.

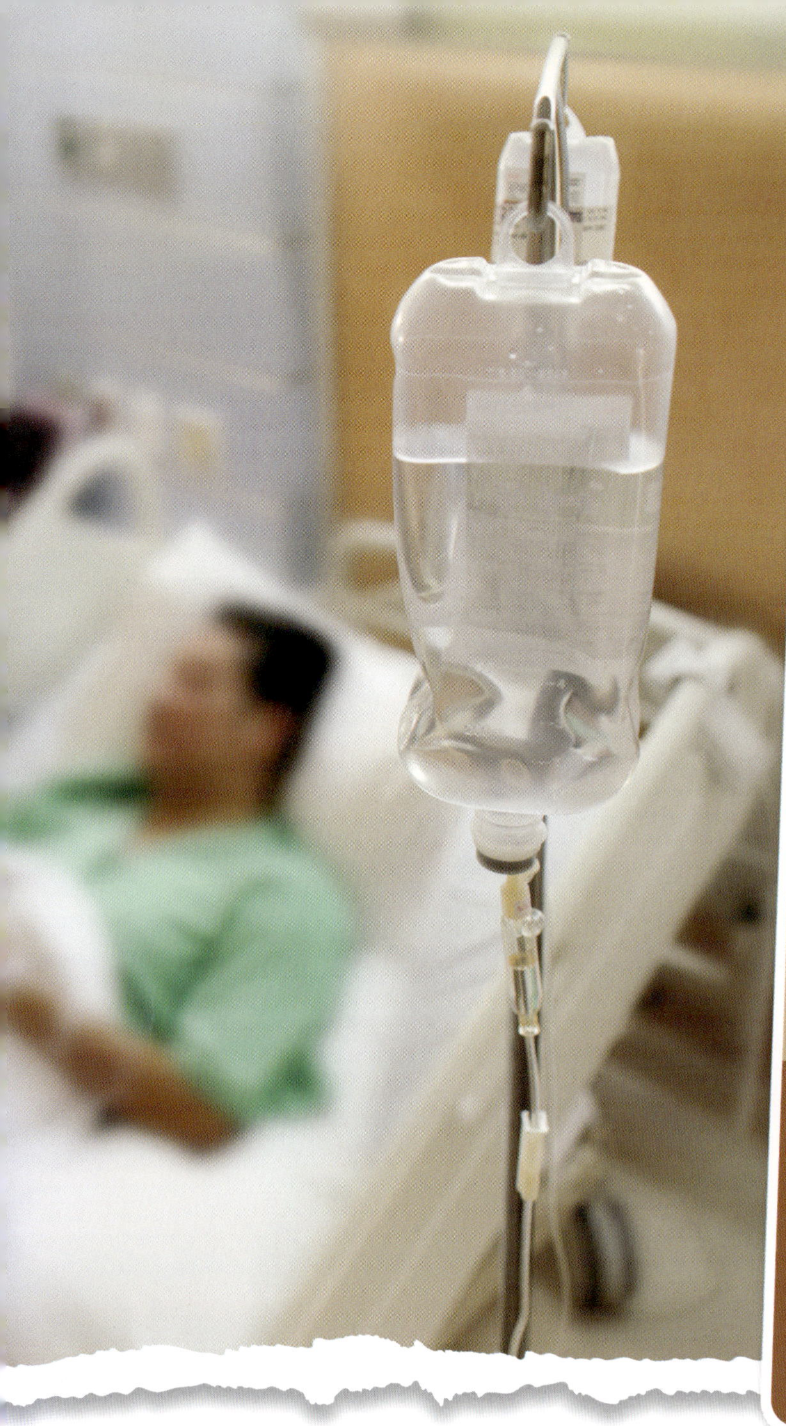

SODIUM CHLORIDE

Chemical Symbol: NaCl

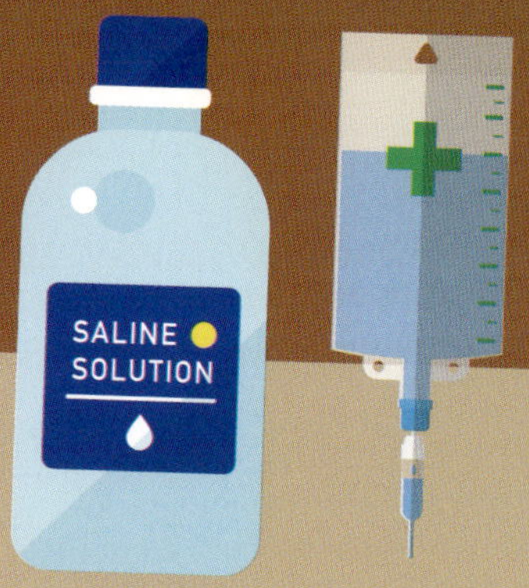

Table salt is a solid that is often mixed with iodine and other agents. It is used in food production, preserving and cooking.

COMMON FORMS

Saline is a sterile, salty liquid that is used in medicine for many purposes, including rehydration, or restoring fluid to the body, and washing wounds. It is also used to soak and clean contact lenses.

USES

Sodium chloride is used in many industries. In the leather industry, it is used for preserving hides, and in the chemical industry, it is used to make other chemicals, such as caustic soda (sodium hydroxide), chlorine and hydrochloric acid.

THE EFFECTS OF SALT

At concentrations of less than 0.9 per cent, salt hydrates, or adds fluid. This is why hospital patients who require more fluid may be given saline intravenously, or through a ‘drip’. At concentrations above 0.9 per cent, salt does the reverse. Rather then hydrating, it dehydrates, or dries out. This is why, at higher concentrations, salt can be used for preserving food.

While seawater contains the highest amount of salt, even water from rivers and dams contains some salt. The root cells of plants are protected by a membrane or barrier that controls the flow of water and salts into and out of the cell. When a plant receives rainwater, it simply absorbs what it needs. When it receives water from other sources, it absorbs only the water, but not the salt it contains. The salt is left behind, and builds up in the soil. This may lead to the soil containing little or no fresh water, and too much salt. If this happens, the plants dry out and die.

WHAT IS SALINITY?

Salinity means 'saltiness', or how salty something is. When the term is used in reference to the environment, it means how much salt is in the soil, waterways and ground water. In Australia, salt has built up in some places to damaging levels. Salinity is now one of Australia's biggest environmental problems. It has already affected at least 2.5 million hectares of farmland, as well as many towns. Salinity has also affected freshwater rivers. This causes serious problems for towns that rely on rivers for their drinking water.

Types of Salinity

Australia is affected by three types of salinity, of these, dryland salinity is by far the most widespread.

- *Dryland salinity, which occurs in rural areas, and is caused by land clearing.*
- *Irrigation salinity, which also occurs in rural areas, and is the result of continued watering, or irrigation, of crops.*
- *Urban salinity, which occurs in cities and towns, caused by a combination of land clearing and over-watering.*

SIGNS OF SALINITY

Salinity is often called the 'white death' because it leaves behind large areas of white crystallised salt. Areas affected by salinity are also often dotted with dead, white tree trunks. Other less dramatic signs of salinity include areas of waterlogged soil in paddocks; and paddocks that are dried-out in places, while remaining green in others. Plant growth is another good indicator of salinity. Plants affected by salinity are slow to grow and look unhealthy. They also suffer from dieback, which is when plants die from the branches or shoots back towards their main trunk or stem.

Cattle and sheep can also indicate if an area is affected by salinity. These animals like some salt in their diet and you might find them licking or digging up the soil in an area affected by salinity.

Salinity may leave its mark on buildings and other structures, such as roads and bridge supports. It can leave white salt traces, which, if left untreated, eat away at these structures, causing them to break up. Salinity can create holes the size of a fist in brick houses.

AUSTRALIA'S SALINITY HOT SPOTS

Every Australian state and territory has a salinity problem. The areas hardest hit include southeastern Western Australia and large areas of the eastern states, including central and western Victoria, southeastern South Australia, and southern New South Wales. The map to the right shows the regions where salinity is an urgent issue.

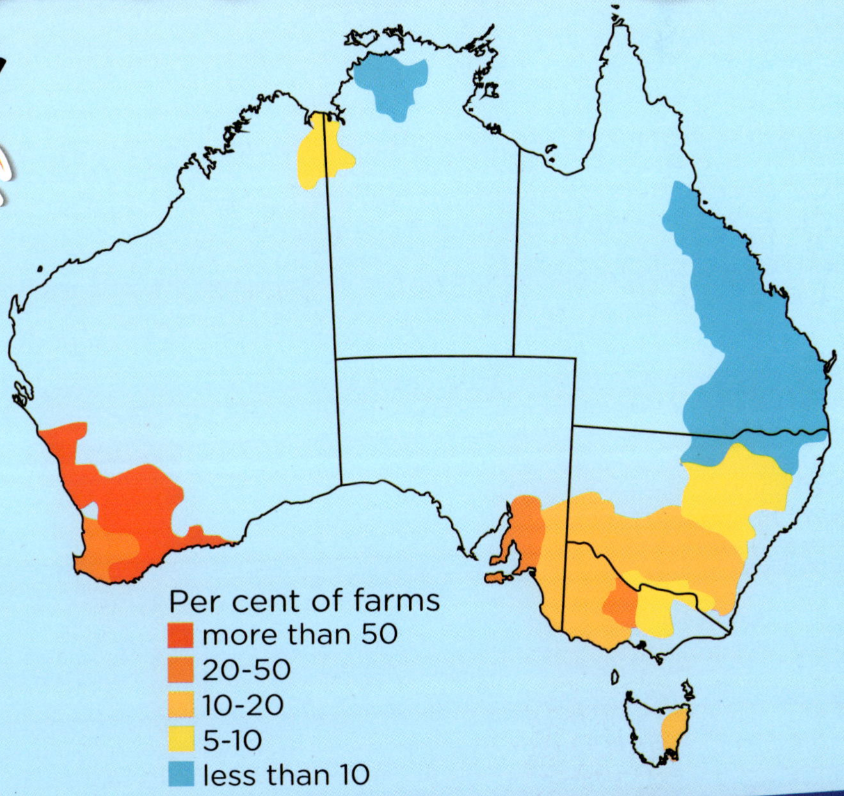

SALINITY SNAPSHOT

01 Around 2 million hectares of agricultural land are affected by salinity.

02 Some 800,000 hectares of this land can no longer be used for agricultural production.

03 Western Australia is the state most affected by salinity, with over 1.2 million hectares affected.

04 Every hour, the Western Australian wheatbelt is losing an area the size of a football field to salinity. About 9 per cent of the wheatbelt's nine million hectares is unproductive because of salt encroachment.

05 In the wheatbelt area salinity has caused a 50 per cent decrease in the numbers of wetland bird species, and 450 plant species are threatened with extinction through salinity.

06

Dryland salinity in the south-west of Western Australia cost farmers an estimated $519 million per year in lost revenue since 2009.

07

Irrigation salinity was first noticed in the 1890s, mainly in the Kerang region of Victoria.

08

In some places, the use of irrigation means there is 100 times more water going into the soil than there was before European settlement.

09

The cost of salinity damage to infrastructure, such as the corrosion and degradation of roads, footpaths, parks, sewage pipes and housing currently costs over $100 million a year.

10

More than 80 country towns across Australia suffer from major salinity issues.

11

Salinity threatens biodiversity through loss of habitat on land and in water.

GROUNDWATER

Groundwater is the water that exists below the ground. It can be absorbed into the top layers of soil where it can be used by plants, stored in underground river systems and basins, and pooled between rock particles deep beneath the Earth's surface.

Importance of Groundwater

Groundwater is essential for the health of soils and plant life. It helps to keep the concentration of minerals and chemicals, including salts, in balance. In some parts of the world, including Africa and central Australia, groundwater is also essential for the wellbeing of humans and their animals. Bores or holes are dug deep into the ground to access clean, fresh drinking water.

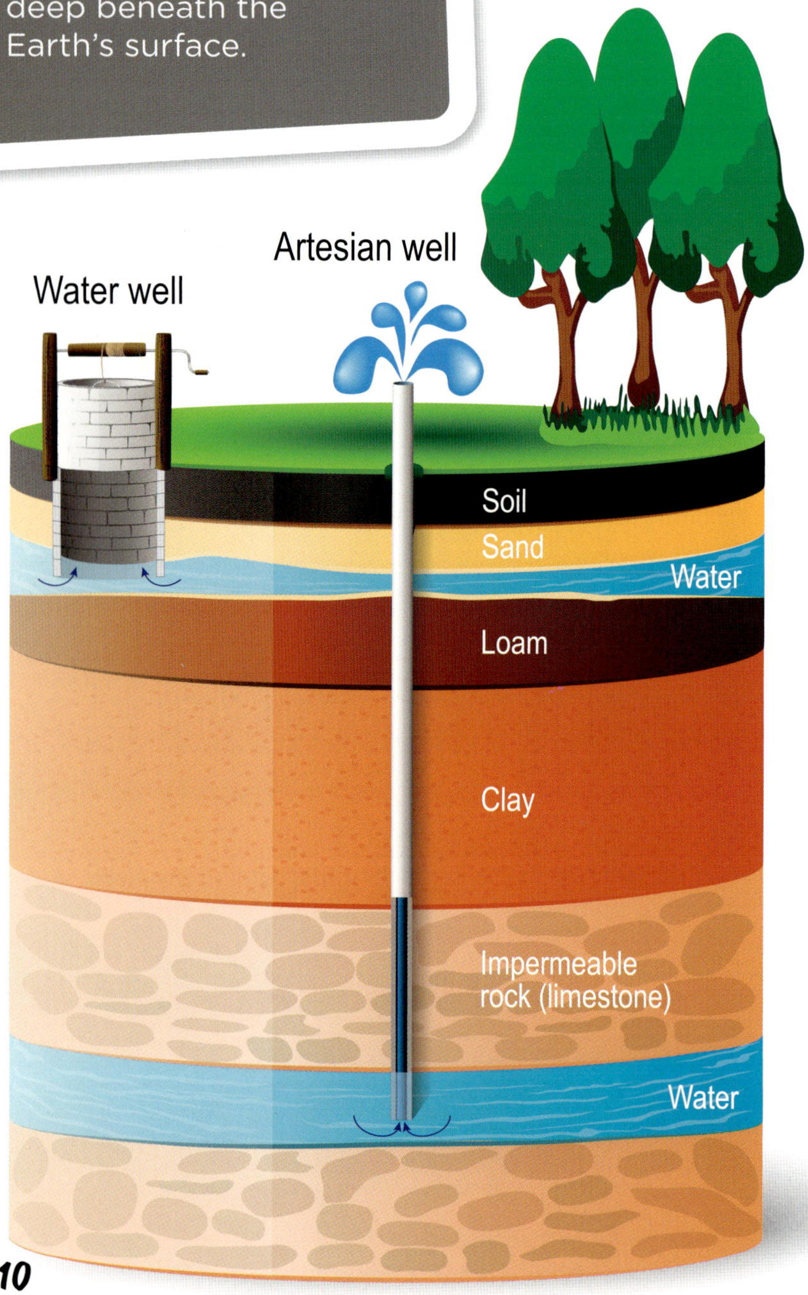

THE WATER TABLE

When water is added to soil, either when it rains or through being watered, whatever is not absorbed into the surface soil seeps further down into the Earth. It eventually hits a layer of rock, and because rock cannot absorb water, the water builds up, and saturates the soil. The area where the soil is completely saturated is called the water table.

The water table rises and drops according to how much water is added to the soil. It rises and drops as the seasons change. During wet months, as more rainwater filters through the ground, the water table rises. In dry, hot months, as ground water evaporates, the water table drops. In the same way, the water table rises and drops depending on how much the land is watered by humans.

How far below ground the water table forms differs greatly from place to place, depending on the type of soil and geographical features. Generally, the water table forms closer to the surface in valleys and low-lying areas. A discharge area is where water pools close to the surface and is lost through evaporation. A recharge area is where water can enter the soil and travel far away from the surface, deep into the Earth.

WATER TABLES AND SALINITY

It is important to understand how water tables rise and drop, it is the rising of water tables that brings the natural salt stored beneath the ground to the soil on the surface. When groundwater that has been carrying salt from deep underground evaporates in a discharge area, it leaves the salt behind in the soil. Salinity is caused by rising water tables. In Australia, water tables are rising because of the way people have changed the natural environment.

The Great Artesian Basin

The Great Artesian Basin flows under Queensland, New South Wales, South Australia and the Northern Territory. It is one of the world's largest underground water reserves, holding an estimated 8,700 million megalitres of water. That is over 350 times more water than in Australia's biggest lake, Lake Eyre in South Australia. Water from the Great Artesian Basin rises to the surface through naturally occurring springs. It can also be accessed by bores dug deep into the ground. The basin has been an essential source of water for irrigation for more than 100 years.

DRYLAND SALINITY

Dryland salinity is caused by deforestation or the widespread culling of trees. When forests are cleared, the land is generally used for farming and is often replanted with food crops or with grasses for cattle.

Native Trees and Shrubs

Australia's native trees and shrubs have helped to keep salinity at bay for millions of years. Many native plants have long, strong roots that penetrate deep into the Earth. Ground water is their main source of water, especially during hot, dry summers and droughts. Most native species are not adversely affected by Australia's naturally salty soils.

When millions of eucalypts and other native species covered Australia, the water table was deep underground. This meant that the salt in the water table also remained deep below the Earth's surface, where it caused little problem.

EUROPEAN FARMING

To support migration and a growing population, European settlers needed to farm crops and livestock. To do so, they cleared vast tracts of land to plant crops such as wheat, which they had grown in their native lands.

Food crops such as wheat are shallow rooted, and so cannot access groundwater to hydrate their root system. Instead, they have to be watered from above. This has a three-fold effect on salinity:

01

First, with all the native vegetation destroyed, there is nothing to drink up the groundwater, so the water table rises. As the water table rises, large amounts of salt are brought to the surface with the water. When the water evaporates, salt is left behind on the surface of the ground. This increases the overall salinity of the soil.

02

Secondly, by clearing all the deep-rooted plants that help to hold the soil together, the land begins to erode. The wind sweeps away the loose, fertile topsoil, leaving behind poorly absorbent clays and other soils. This also causes the water table to rise.

03

Thirdly, the constant watering of crops means that more and more water is pumped into the ground. This in turn causes irrigation salinity. The plants absorb what water they require, leaving the salt behind in the soil. This causes a build-up of salt and eventually the soil becomes too salty for crops to grow.

HOPE FOR THE FUTURE

One possible solution to salinity is biochar, a charcoal produced from plant matter. After positive results in the reduction of salinity on a small farm in Brookton, Western Australia, companies and researchers are undertaking official trials on biochar and its impact on salinity.

DRYLAND SALINITY

This diagram shows how the water cycle and land clearing affects the level of groundwater and the amount of salt in the soil.

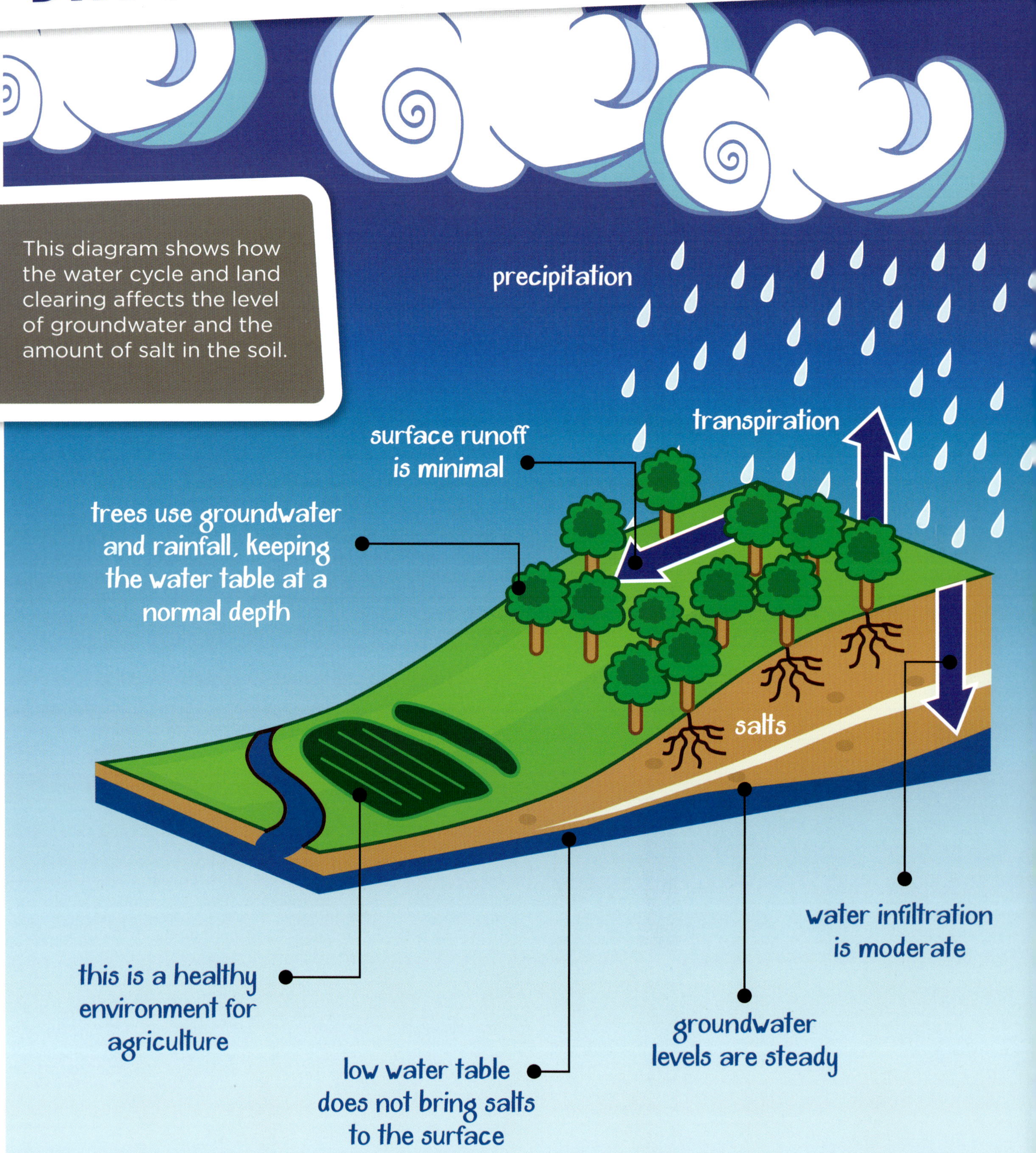

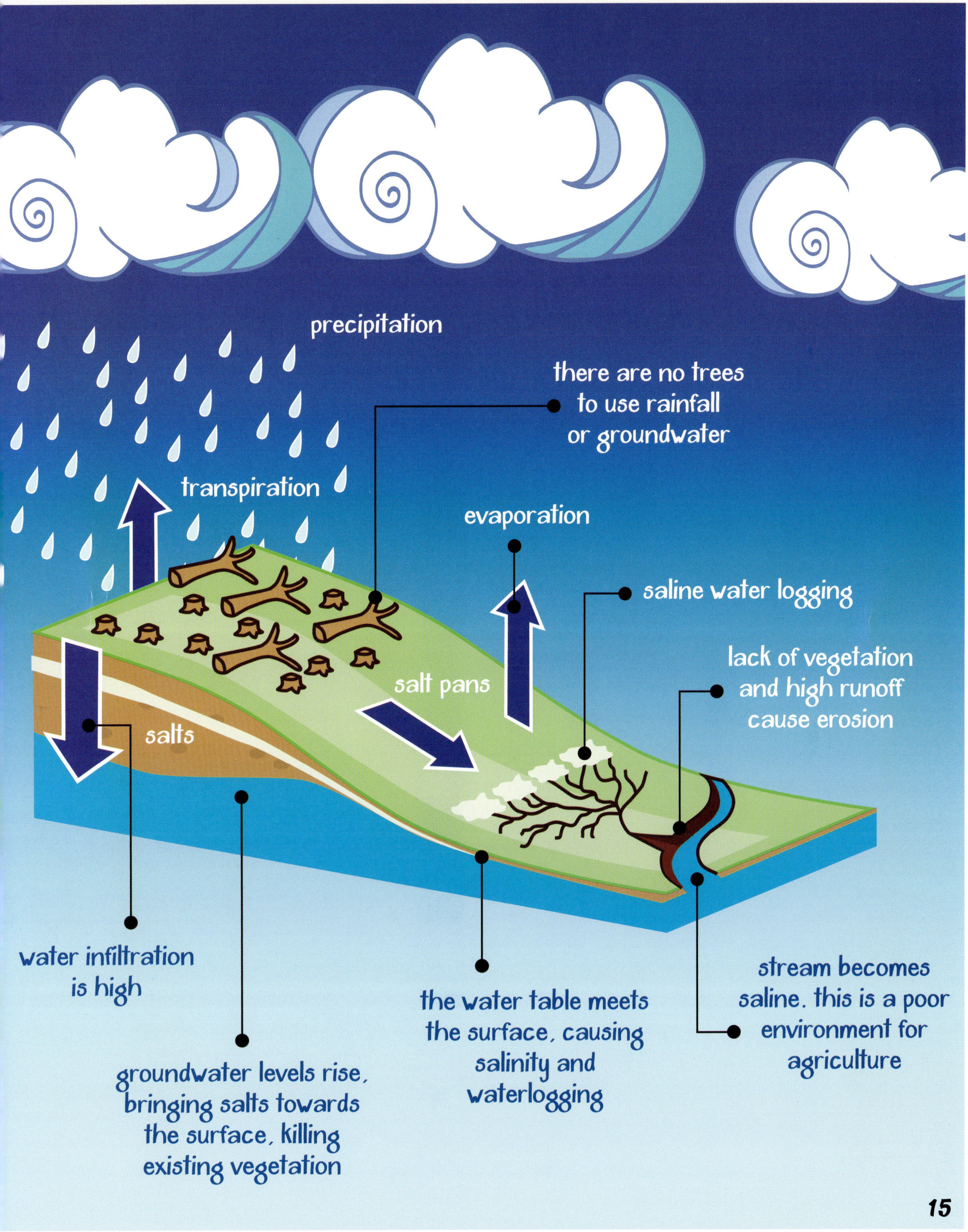
precipitation
there are no trees to use rainfall or groundwater
transpiration
evaporation
saline water logging
lack of vegetation and high runoff cause erosion
salt pans
salts
water infiltration is high
groundwater levels rise, bringing salts towards the surface, killing existing vegetation
the water table meets the surface, causing salinity and waterlogging
stream becomes saline. this is a poor environment for agriculture

Dryland Salinity Case Study

The Parsons, South Australia

Two South Australian farmers, Wolford and Marie Parsons, dedicated much of their lives to fighting local dryland salinity. In 2006, their hard work was recognised with two major awards: the Rural Press Landcare Primary Producer Award and the Individual Landcare Award.

The Parsons purchased a 6.3-square kilometre mixed food crop and livestock farm in Point Vincent on South Australia's Yorke Peninsula. At that time, the farm, called 'The Springs', was in environmental ruin. Within a 20-year period from the late 1940s, more than 95 per cent of the original native vegetation had been cleared for farming. Although Marie Parsons managed to save the last remaining 0.28 square kilometres of scrubland and lagoons, the damage was already done.

The first signs that the area was suffering from dryland salinity were seen after widespread flooding in 1979, which affected a large portion of 'The Springs'. When the flood waters receded, Marie and Wolford discovered that more than 80 hectares of the farm was covered in a crusty layer of salt. Over time, the problem spread to the rest of the farm, making crop production almost impossible. If the Parsons did not take urgent action, they would lose their home and their livelihood.

In 1982, the Parsons set about draining the salty water out of the worst affected areas into a newly created salt lake. They also stopped planting these areas with shallow-rooted food crops. The next step was to revegetate the land. Over the next six years, the Parsons planted some 150,000 saltbush plants, as well as a variety of salt-tolerant grasses and animal fodder crops. By 2003, 'The Springs' was being run as a grazing, or livestock, farm. The only crops grown were lucerne and other 'water table friendly' animal fodder crops. The Parsons also adopted new, environmentally friendly farming techniques, including planting pastures according to their soil type.

Wolford and Marie Parsons are now dedicated to sharing their experiences and knowledge with the wider community. They strongly believe that everyone needs to work together to fight salinity. They have been involved in research projects at the University of Adelaide and have given talks to the South Australian Parliament and the Commonwealth Scientific and Industrial Research Organisation (CSIRO), Australia's leading scientific and research organisation.

Today, 'The Springs ' is in far better shape than it was decades ago. The Parsons have planted over 350,000 trees on the property. Many native animals have returned to the area, and native plant species are thriving there once again.

IRRIGATION SALINITY

Irrigation salinity is caused mainly by the over-watering of crops. When more water is applied than plants can use, the excess seeps downward, adding to the groundwater and causing the water table to rise. When the saline water table rises to within 2 metres of the surface, more water evaporates, leaving behind the salt. The layer of salt prevents further evaporation, so the soil beneath it becomes waterlogged. Waterlogged soil contains too much water and not enough oxygen. Plants die in waterlogged soil because they cannot get enough oxygen, which, like water, they absorb through their roots.

Additionally, plants are able to absorb water from the soil, without the salt it contains. This leaves a further build-up of salt behind. Eventually, there is too much salt in the soil for plants to grow. This is made worse by heavy rains, but drought improves the situation. Often, the only water available for irrigation is saline bore or groundwater, or the water comes from salty rivers, which further increases the risk of salinisation. Irrigation salinity is responsible for only about 7 per cent of Australia's salinity problem. Nevertheless, climate change could exacerbate the problem. During long periods of drought, farmers need to water their crops more regularly, and long, hot summers dry out the ground, making it hard and water repellent, again increasing the amount of water farmers must use to reach plants' root level.

Irrigation Salinity Case Study

Western Australia's Coastal Plains and Wetlands

The Swan and Scott coastal plains in Western Australia have poorly drained soils and shallow, naturally saline groundwater. This makes them very susceptible to irrigation salinity, particularly as use of irrigation increases on farms in the area. The Western Australian Government has predicted that irrigation salinity will become a severe problem in up to 40 per cent of the Swan Coastal Plain, from Gingin to Dunsborough.

The Ord River Irrigation Area is a major horticultural area that is surrounded by wilderness. It is also under threat of irrigation salinity. To try to slow the rise of the water table, dewatering bores are now required in some areas. These bores pump excess groundwater from the ground into large, above-ground storage areas.

Many of Western Australia's wetlands are also feeling the impact of irrigation salinity. Increasing salinity has caused substantial losses of native vegetation around some wetlands, which in turn has caused a fall in waterbird populations, particularly in the Coyrecup, Coomelberrup, Walymouring, Eganu, Dumbleyung and Parkeyerring wetlands.

URBAN SALINITY

Urban salinity occurs in built-up areas - places where there are lots of buildings, roads and people. Urban salinity is a growing problem in Australia and is starting to have a major impact on the infrastructure of many towns and cities. Below all the concrete, bricks and roads lie the ground and its water table. As it does elsewhere, this water table rises and falls with the seasons, but also because of human activity.

As Australia's population grows, more land is cleared for houses, schools, shops, roads and other facilities, and all this leads to a form of dryland salinity. As well as clearing trees (thus removing the deep-root systems of plants that would absorb a lot of water), we tend to use a lot more water in urban areas.

Home gardens in Australia tend to have few native trees and plants, but lots of water-loving, shallow-rooted introduced plants. The regular watering these gardens require increases the amount of water in the ground, leading to a form of irrigation salinity.

The buildings and roads in urban areas contribute to the amount of water entering the ground by creating runoff from their roofs and gutters when it rains. There are also underground water pipes that sometimes leak. All this extra water in urban areas causes the water table to rise, bringing the salts in the soil to the surface.

The National Land and Water Resources Audit (2001) predicted that 67,000 kilometres of road, 5,100 kilometres of railways and 220 towns are at high risk of damage due to urban salinity by 2050. In 2018, there has been little change to this.

Urban Salinity Case Study

Bendigo

The City of Greater Bendigo, in central Victoria, was established in 1851 when gold was discovered in the area. With a population of more than 95,000, it is now Victoria's third-largest city. Urban salinity problems have been occurring in Bendigo since the 1980s. These problems occur most often in the lower-lying areas of Bendigo's landscape, like valley floors and the floodplains of creeks, as this is where the water table lies closest to the land's surface.

As in many other areas, it took people a while to realise that salinity was not just a rural problem, but an urban one, too. In Bendigo, urban salinity has damaged buildings, roads, bridges, pipelines, parks, lakes and private homes and gardens. The damage caused by salinity can be very expensive to repair.

The people of Bendigo developed systems to help manage this problem. They identified the most saline areas of the city so future building and development could be planned around them. They have been monitoring major infrastructure (for example, roads and important buildings) for signs of salinity damage. By doing this, the people of Bendigo aim to prevent salinity damage to new areas and identify existing salinity damage before it becomes too great.

The signs of urban salinity include:

- Waterlogged or boggy soils.
- Patches of no, or poor plant growth.
- Salty white powder on brickwork, roads, footpaths and other structures.
- Cracked and crumbling roads and footpaths.
- Corroded underground pipes.

THE MURRAY-DARLING BASIN

The Murray-Darling Basin is an enormous area of 1,060,000 square kilometres that runs across Queensland, New South Wales, the Australian Capital Territory, Victoria and South Australia. It encompasses the river system of the Darling, Lachlan, Murray and Murrumbidgee rivers and the areas that surround them. This important area provides water to three million people. The Murray and Darling rivers are the lifeblood of the eastern states of Australia and South Australia. The Murray River is Australia's most important source of fresh water.

The Murray-Darling Basin is a naturally saline environment. In 1829, Charles Sturt (the first European to discover the Darling River) recorded that he could not drink the water in the Darling because it was too salty. The salts come from ground water, the weathering of rocks and precipitation.

The Murray-Darling Basin is affected by both irrigation and dryland salinity. Over the past 100 years, at least 15 billion trees have been cleared from the Murray-Darling Basin. As a result of this, groundwater levels have been rising, increasing the salinity of the rivers in the basin.

The Murray River is a regulated river, which means that people control how the river flows. During the 1920s, dams and weirs were built along the river to control the river's flow by storing and releasing its water.

Almost 80 per cent of the river's flow is used by people, mostly for irrigation. It produces 50 per cent of Australia's irrigated produce. This has two effects on salinity. First, the water that is diverted from the river for irrigation is redistributing salt across the landscape. Secondly, diverting water away from the river causes the Murray to be in drought more often than it would be under natural conditions. When the river is in drought, the flows of water in it are very low. There is a clear relationship between river flow and salinity levels, the lower the river flow, the higher the level of salt.

Over the next couple of decades, at current levels, salinity will increase to a point where water from the Murray-Darling Basin will be outside World Health Organization recommended levels for drinking water for much of the year.

The lower Avon-Richardson catchment is one of the most severely salt-affected areas of central north Victoria. The problem was first noticed in the 1960s, when farmers began reporting diminished crop productivity. Some parts of the Richardson River are four times more saline than seawater.

Adelaide draws about 55 per cent of its water supply from the Murray River in a normal year, and up to 90 per cent during a drought.

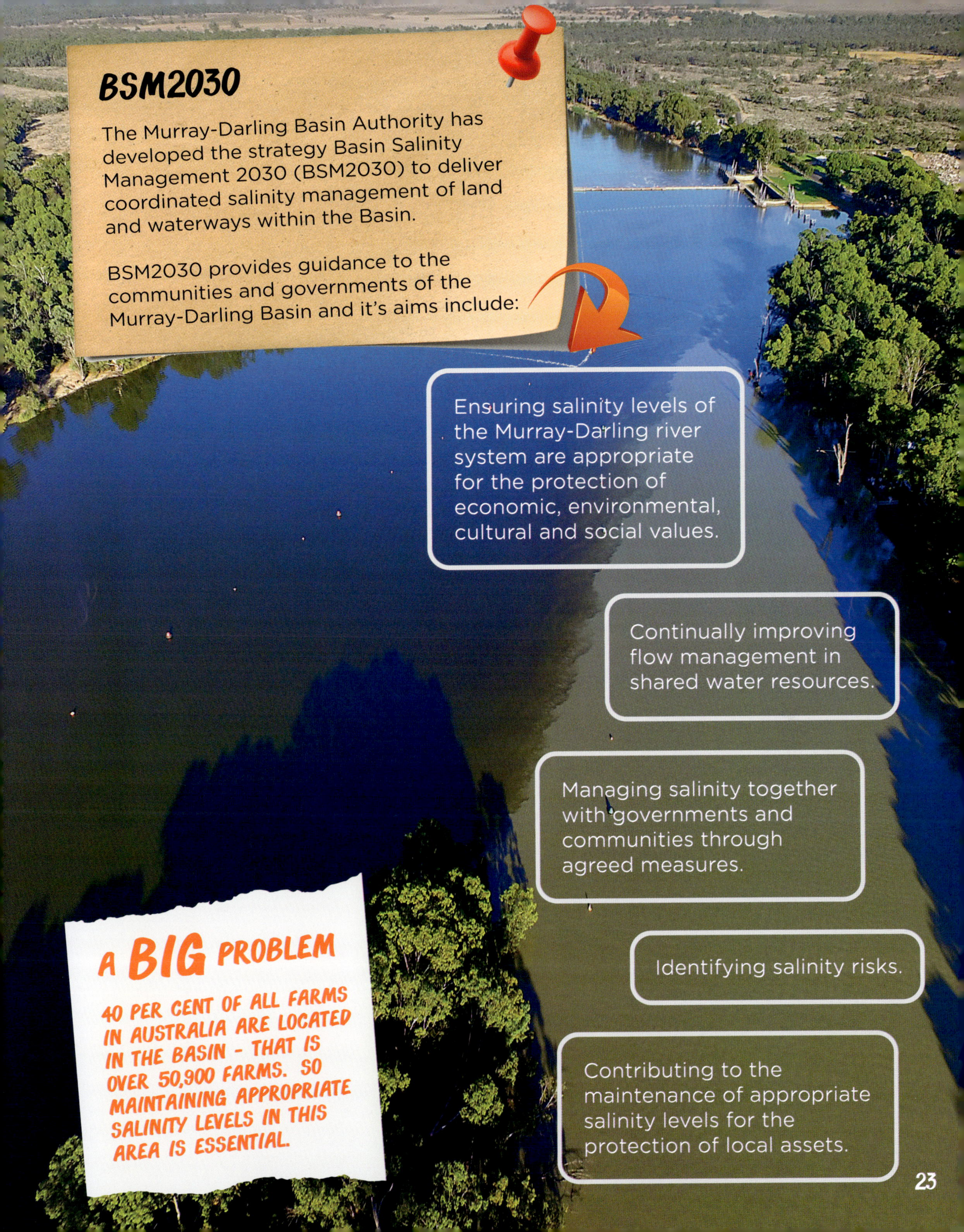

BSM2030

The Murray-Darling Basin Authority has developed the strategy Basin Salinity Management 2030 (BSM2030) to deliver coordinated salinity management of land and waterways within the Basin.

BSM2030 provides guidance to the communities and governments of the Murray-Darling Basin and it's aims include:

- Ensuring salinity levels of the Murray-Darling river system are appropriate for the protection of economic, environmental, cultural and social values.
- Continually improving flow management in shared water resources.
- Managing salinity together with governments and communities through agreed measures.
- Identifying salinity risks.
- Contributing to the maintenance of appropriate salinity levels for the protection of local assets.

A BIG PROBLEM

40 PER CENT OF ALL FARMS IN AUSTRALIA ARE LOCATED IN THE BASIN - THAT IS OVER 50,900 FARMS. SO MAINTAINING APPROPRIATE SALINITY LEVELS IN THIS AREA IS ESSENTIAL.

SALINITY AND WATER QUALITY

Salinity levels have a huge impact on the quality of water available for drinking, domestic, industrial and agricultural use. This is because when saline groundwater rises it does not just affect the land, it flows into rivers and other water sources as well. Groundwater that runs into surface water is called baseflow.

Salt that has been brought to the Earth's surface by groundwater is also washed into rivers when it rains. This is called washoff. In areas that use irrigation, leftover water from irrigation can also drain back into rivers, carrying salt from the ground. All this further increases the salinity levels of the rivers. Throughout Australia, organisations have been formed to combat water salinity.

Water Authorities

The North Central Catchment Management Authority (NCCMA) and the Murray-Darling Basin Authority (MDBA) are two of the many bodies formed to combat water salinity. The NCCMA is dedicated to improving the condition of Victoria's waterways, including the Avoca, Avon, Campaspe, Murray and Richardson rivers. The MDBA is dedicated to saving the Murray and Darling rivers from salinity and other environmental problems. It is a partnership between the South Australian, Victorian, Australian Capital Territory, New South Wales and Queensland governments and the federal government.

PLANTING FOR THE FUTURE

The best way to plan for a salinity-free future is to plant for the future.

Reclaiming Lost Land

It is important to revegetate land that has lost its plant life through salinity. Many native plants have long roots, which bind the soil together and so help prevent erosion, or the loss of surface soil to rains, wind and other environmental causes. Revegetation helps make the soil more fertile. When leaves fall to the ground, they are composted back into the Earth, providing nutrients for new plants. Reclaiming lost land can also help establish new practices such as saltbush farming.

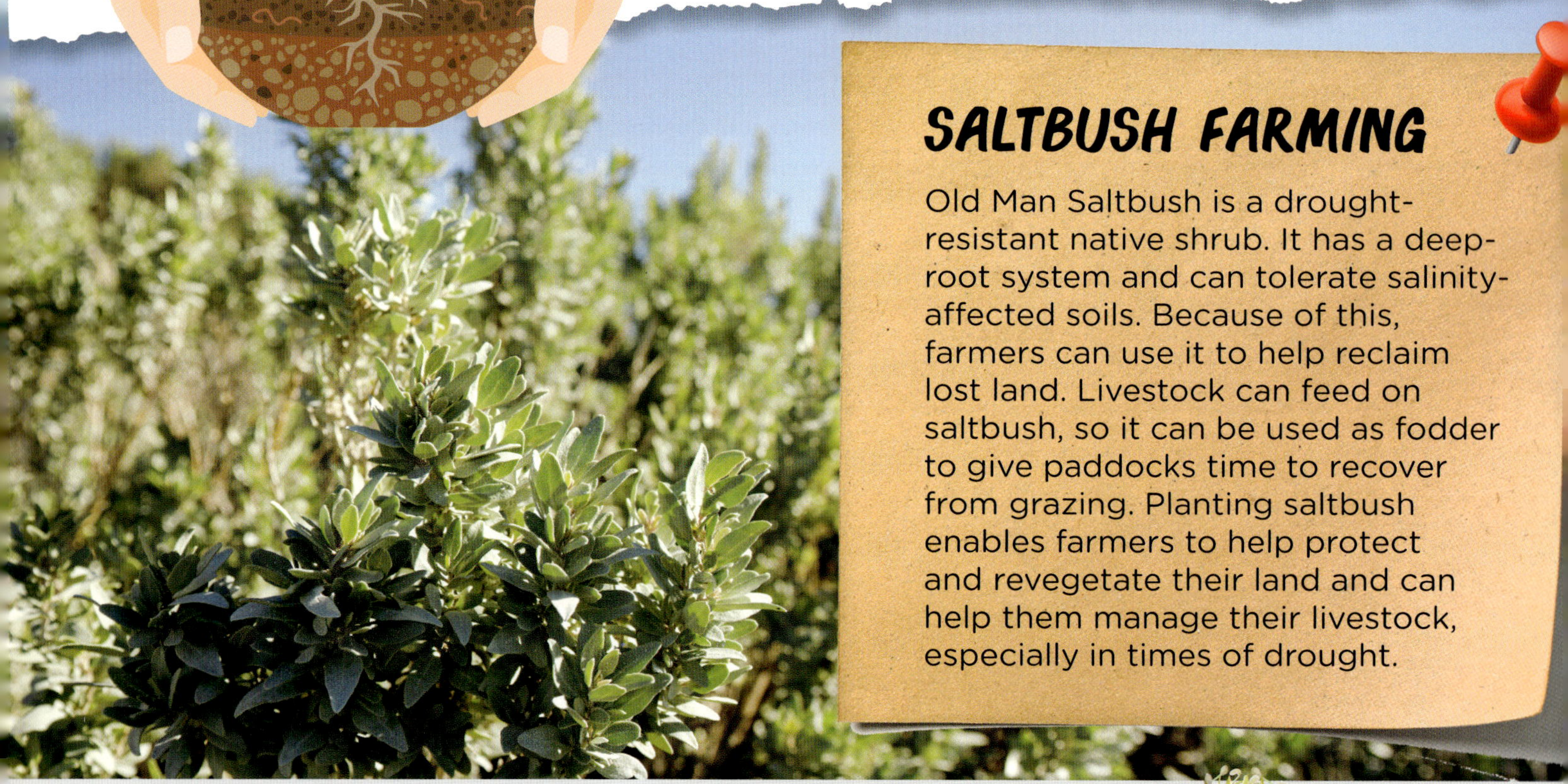

SALTBUSH FARMING

Old Man Saltbush is a drought-resistant native shrub. It has a deep-root system and can tolerate salinity-affected soils. Because of this, farmers can use it to help reclaim lost land. Livestock can feed on saltbush, so it can be used as fodder to give paddocks time to recover from grazing. Planting saltbush enables farmers to help protect and revegetate their land and can help them manage their livestock, especially in times of drought.

PROTECTING SALINITY-FREE LANDS

Revegetating salinity-free areas with native plants is a good way to stop the spread of dryland salinity. Deep-rooted plants help to lower the water table by sucking up groundwater. They also help protect the soil from erosion. Additionally, native plants help support Australia's unique animal life. Many birds, insects, reptiles and other animals depend on native plants for food, shelter and even reproduction.

SMART CROP FARMING

Salinity can be stopped by improved irrigation practices. These include supplying irrigation water through pipes instead of open channels. In pipes, the water cannot evaporate or leak into the ground. Another way to control salinity is to use less water when irrigating crops; sprinkler and drip systems use less water than other types of irrigation. Now, farmers also have tensiometers, which are instruments that measure how hard a plant is working to extract moisture from the soil. This reading enables farmers to gauge how much water a crop will require and how much irrigated water is therefore needed. This helps prevent over-watering of crops, the excess water from which runs straight into the ground.

Salinity can also be stopped by switching to more salt-tolerant, 'water table friendly' crops such as lucerne, vetiver grass, wheat grass and puccinellia. It is also important that farmers do not leave their fields fallow for long periods of time, as this allows more water to enter the ground.

Tensiometer

HOW YOU CAN HELP COMBAT SALINITY

Encourage people to water their gardens wisely. We should always preserve water, even when dams are full and there are no water restrictions. This stops over-watering and so helps to keep the water table low and prevent salinity.

Encourage your family, friends and neighbours to plant more native trees and plants in their home gardens. The natives, with their longer roots, are able to take their water from the water table instead of having to be watered. This helps to keep the water table low and prevent salinity.

Take a look around your local area to see if there are any creeks, rivers, parks, paddocks or vacant council land that could do with some tender loving care. Contact your local Landcare group and ask them to organise a community action day to clean up and revegetate the area. If there is not a Landcare group in your area, see if you can start one yourself!

Write to your local council and tell them you would like them to be more salt conscious. Ask them to make sure that native vegetation is planted in any parks or gardens they set up and that native plants are included in the plans for new building projects.

Ask your school principal if any ugly or under-used parts of the school ground can be converted into native gardens. You could also ask for student volunteers to make the gardens, if permitted. Consider ways of letting others know why the project is being undertaken.

Read up on the types of native plants that are suitable for your area so that next time you go to a garden nursery with your family or friends, you can tell them about the beautiful natives they could buy instead of more traditional imported plants.

Lobby your local council to replace old terracotta or metal underground pipes with plastic piping. Doing this helps stop water leakage through broken or corroded pipes.

COMBATING SALINITY

The Coobabla Saltbush Project

The Coobabla Saltbush Project in Western Australia manages salinity by planting saltbush, which helps reduce salinity. Over 9,000 saltbush seedlings have been planted.

Throughout Australia, many groups, including the federal, state and territory governments, have come together to combat salinity. This has led to the implementation of various audits and action plans. One of the most effective groups is Landcare Australia.

Landcare Australia

Landcare Australia is a national organisation whose primary purpose is to care for all of Australia's land, whether it is bush, farming land or urban land. It is a partnership between urban and country communities, governments, businesses and other organisations. There are now over 5,400 Landcare and Coastcare groups across the Australia.

Landcare and its partners organise activities such as tree-planting days and river clean-ups, as well as training and information sessions. Through these activities, Australia's precious natural resources are being repaired and viably managed.

Landcare Australia

The National Landcare Program is a key part of the Australian Government's commitment to protect and conserve Australia's water, soil, plants, animals and ecosystems. The Department of Agriculture and the Department of the Environment have provided funding to Landcare.

GLOSSARY

audit - an official examination of records and statements

baseflow - groundwater that runs into surface water

bore - a large hole made down into groundwater

catchment - an area that water from dams and rivers drains into

climate change - the process by which overall temperature moves from one average to a new average

corroded - occurs when a substance gradually eats away at a surface

crystallised - taking the form of crystals

deforestation - the widespread chopping down of trees and forest in order to create farmland or land for humans

dehydrate - to lose or remove water

discharge - area where water pools close to Earth's surface and is lost through evaporation

drought - a long period of little or no rainfall

erosion - the wearing away of land surface by wind, water and sun

evaporation - the process by which water becomes vapour

fallow - empty, not planted with seeds

fodder - food given to livestock

groundwater - the water that is found beneath the surface of the land

hydrate - to add fluid

imbalance - the state or condition of lacking balance

indicator - something that shows or gives information

infrastructure - an organised system of roads and buildings

intravenously - the act of injecting saline or blood into a vein

irrigation - extra watering above and beyond rainfall

monitoring - controlling or checking a process

native - a plant or animal living in its country of origin

precipitation - moisture falling from the sky in the form of rain, snow or hail

productivity - a measure of how efficient an activity or process is

recharge - area where water can enter the soil and travel far away from the Earth's surface

revegetate - the process of planting plants to restore the plant life of cleared areas

runoff - excess water from rainfall that flows across the land

rural - having to do with farming or countryside

saturate - to wet thoroughly

sediment - matter (dirt and other particles) that settle at the bottom of a river or body of water

toxic - poisonous

washoff - salt that has been brought to the Earth's surface by groundwater and is washed into rivers when it rains

waterlogged - saturated with water

weir - a barrier across a river or stream that is used to collect water and prevent it from going downriver

wetlands - a place that is covered with water some or all of the time

INDEX

FURTHER INFORMATION

Using the Internet

Explore the Internet to find out more about salinity, the ground water table and the ways to combat salinity as mentioned in this book. Keywords include 'dryland salinity', 'rising water table' and the names of people, places and events you are interested in.

https://www.mdba.gov.au/managing-water/salinity
http://www.agriculture.gov.au/ag-farm-food/natural-resources/salinity